THE ULTIMATE
CISSP STUDY PLAN

A comprehensive guide to success

Dr. Leonard Simon

CONTENTS

INTRODUCTION

UNLOCKING THE POWER OF CISSP CERTIFICATION

In the fast-changing digital world of today, there has never been a greater demand for talented and informed cybersecurity workers. The Certified Information Systems Security Professional (CISSP) credential has become the gold standard in the area as enterprises around the world try to protect their sensitive data and keep the integrity of their systems. This internationally recognized certification not only proves a professional's knowledge and skill in information security, but it also helps them go up in their careers and improve as people.

The International Information System Security Certification Consortium (ISC2) is in charge of the CISSP certification, which is widely valued by both companies and leaders in the field. Getting this coveted certification shows that you have a comprehensive understanding of the latest cybersecurity principles, methods, and technologies, as well as the ability to design, implement, and maintain a secure information system environment. It's no surprise that people with CISSP certification are in great demand and often get paid more than others who don't have certification.

This thorough guide is meant to help you get the CISSP certification by giving you an effective study plan that fits your specific needs and learning style. By using the ideas, tactics, and resources in this eBook, you will not only be better prepared to pass the difficult CISSP exam, but you will also develop the confidence and abilities you need to do well in your cybersecurity profession.

Welcome to the path to becoming a Certified Information Systems Security Professional. This is where your path to success begins.

NAVIGATING THE EIGHT DOMAINS OF THE CISSP EXAM

The Certified Information Systems Security Professional (CISSP) exam is meant to test how well you know and understand a wide range of information security principles and methods. To do well on the test, you need to know a lot about the (ISC)^2 CISSP Common Body of Knowledge (CBK), which is made up of eight domains. These domains include the essential skills that every cybersecurity expert should have. Together, they make up a complete framework for securing the information systems of a business.

In this eBook, we'll go through the eight CISSP domains and provide you with a deep grasp of the issues covered in each, as well as the key concepts, technologies, and best practices you need to know. The domains are:

1. **Security and Risk Management:** This domain covers essential principles of security, risk management, and compliance, including legal, regulatory, and ethical considerations, as well as security policies, standards, and procedures.

2. **Asset Security:** This domain focuses on the protection and management of information assets, including data classification, ownership, and handling, as well as privacy protection and data retention.

3. **Security Architecture and Engineering:** This domain encompasses the design and implementation of secure information systems, including cryptography, secure network architecture, and system security engineering principles.

4. **Communication and Network Security:** This domain covers the fundamentals of secure network communication, including network architecture, protocols, segmentation, and secure design principles.

5. **Identity and Access Management (IAM):** This domain deals with the management of user identities and access controls, including authentication, authorization, and accountability mechanisms, as well as identity and access provisioning lifecycle.

6. **Security Assessment and Testing:** This domain focuses on the methodologies and techniques used to evaluate the effectiveness of security controls, including vulnerability assessments, penetration testing, and security audits.

7. **Security Operations:** This domain covers the day-to-day management of security operations, including incident response, disaster recovery, business continuity planning, and the secure administration of information systems.

8. **Software Development Security:** This domain addresses the integration of security best practices throughout the software development lifecycle, including secure coding, application security, and DevSecOps.

As you read through the eBook, you will learn more about each area and get the skills and information you need to do well on the CISSP exam. By learning everything you can about these eight topics, you will develop a strong foundation for your cybersecurity career and make getting the highly sought-after CISSP certification easier.

CRAFTING YOUR PERSONALIZED CISSP STUDY PLAN

It might be challenging and take a lot of time to become a Certified Information Systems Security Professional (CISSP). Since there is a lot of information to learn and understand, studying in a methodical and structured way is essential. A well-thought-out study strategy is your road map to success. It shows you how to get through the difficult CISSP exam domains and ensures you use your study time wisely.

This eBook will help you make a customized CISSP study plan that fits your schedule, learning style, and knowledge base. By creating a personalized study plan, you can make the most of your time and resources and ensure you cover all the crucial subjects and ideas you need to know for the CISSP exam. This targeted approach will help you pass the exam but will also help you learn more about the subject, which will help you do well in your cybersecurity job.

In this eBook, you'll find helpful ideas, suggestions, and methods for your study journey. We will provide the tools and techniques to make a study plan that works for you, from figuring out what you already know and how much experience you have to set realistic goals and deadlines. We will also talk about different ways to study, organize your time, stay motivated, and deal with problems along the road.

By following the study plan in this eBook, you will be better prepared to confidently take the CISSP exam and get the prestigious certification that can open doors and take your career to new heights. So, let's start this exciting journey together and take the first step toward CISSP success.

1

GETTING STARTED

ASSESSING YOUR CURRENT KNOWLEDGE AND EXPERIENCE

Before you start the CISSP study plan, you need to look at what you already know and have done in the field of information security. This test will give you a place to start and help you figure out where to focus your efforts and how to use your study time. Here are some steps to help you assess your current knowledge and experience effectively:

1. **Review the CISSP exam domains:** Begin by familiarizing yourself with the eight domains of the CISSP Common Body of Knowledge (CBK). Take note of the specific topics covered in each domain, as well as any areas that seem particularly challenging or unfamiliar to you.

2. **Reflect on your professional experience:** Consider your work history and the extent to which you have been exposed to the various CISSP domains. Have you been involved in security risk management, network security, or software development security? Make a list of the areas in which you have hands-on experience and those where your exposure is limited.

3. **Take a self-assessment quiz or practice test:** To gauge your current knowledge, take a preliminary self-assessment quiz or practice test covering the CISSP domains. This exercise will provide valuable insights into your strengths and weaknesses, helping you identify topics where you need to invest more study time.

4. **Analyze your results:** After completing the self-assessment or practice test, review your results carefully. Look for patterns in your performance – are there specific domains or topics where you consistently struggle? These areas may require more focused attention during your study sessions.

5. **Create a knowledge and experience map:** Based on your analysis, create a visual representation of your knowledge and experience across the CISSP domains. This map can be as simple as a table or chart, with columns for each domain and rows for various topics. Fill in the cells with your level of expertise, using a color-coded system or numerical rating scale.

6. **Set realistic expectations:** Keep in mind that the CISSP certification is designed for experienced professionals, so it is natural to encounter unfamiliar concepts and areas where your knowledge may be lacking. Be honest with yourself about your current knowledge and experience, and set realistic expectations for your study journey.

Looking honestly at what you already know and have done, you may make a focused and successful CISSP study strategy that fits your demands. This customized strategy will ensure you spend the time and use the right tools to learn the most crucial information security principles. This will help you pass the CISSP exam.

DETERMINING YOUR LEARNING STYLE

For making a good CISSP study plan, knowing how you learn best is essential. By adapting how you study to how you naturally learn, you can improve how much you remember, understand, and love learning in general. Here are the steps to help you determine your learning style and leverage it to optimize your CISSP exam preparation:

- **Check out how people usually learn:** Learn about the most common learning styles, which are:
  - **Visual:** People who learn better through pictures, diagrams, and other visual aids do so.
  - **Auditory:** People who learn best by hearing like to listen to lectures, dialogues, and audio recordings.
  - **Read/Write:** People who are good at reading and writing are good at learning through things like textbooks, articles, and taking notes.
  - **Kinesthetic:** People who learn best via touch, movement, and real-world experiences are kinesthetic learners.

- **Think about what you want:** Think about what you've learned in the past and which strategies have worked best for you. Think about the times when you were most interested and could remember things. This self-test can help you figure out what your primary way of learning is.

- **Take a quiz about how you learn:** Several online quizzes might help you determine how to study best. Most of the time, these tests are made up of questions to decide how you like to learn and your primary learning style.

- **Combine different ways of learning:** Remember that most people learn best when using a mix of other study methods. Even though you may have a dominating learning style, using parts of different styles might improve your entire study experience and help you remember essential ideas.

- **Customize your study strategy:** Once you know how you learn best, you can create your CISSP study plan to fit that. For example, if you are a visual learner, show crucial ideas with diagrams, charts, and mind maps. If you learn best by hearing, use podcasts, taped lectures, and discussion groups as part of your study routine.

- **Experiment and adapt:** Be willing to try different learning methods as you study. Pay attention to what works best for you, and be ready to change your study plan as needed to get the most out of your learning.

You may make studying for the CISSP exam more manageable and enjoyable by discovering how you learn best and changing your study methods to match. This personalized study plan will not only consider how you know best but also help you retain and utilize the information you need to do well on the CISSP exam and in your cybersecurity career.

SETTING REALISTIC GOALS AND DEADLINES

Setting realistic goals and dates is vital to developing a solid CISSP study plan. By setting goals and deadlines that you can attain, you can stay motivated, keep track of your progress, and stay on track as you study for your exam. Here are some tips to help you set realistic goals and deadlines for your CISSP studies:

- **Look at your existing situation.** Think about your work and personal obligations and how much time you can study for the CISSP exam. Be honest about how much time you have to study, and consider any interruptions or busy times that might alter your timetable.

- **Split the CISSP domains up:** Break up the eight CISSP test domains into smaller subtopics that are easier to handle. This method will help you focus on specific areas of study and keep better track of your progress.

- **Make short-term and long-term goals:** Set short-term and long-term goals for yourself while you prepare for the CISSP exam. Short-term goals can be to master a specific topic, finish a chapter in a study guide, or earn a particular score on a practice test. Long-term goals could include completing all the study materials, becoming very good at all the CISSP domains, or setting up the exam.

- **Define priorities based on your knowledge gaps.** When you define your goals, give the topics and domains where you found knowledge gaps the most weight. Give these areas more time and money because they will need more work and focus to master.

- **Make a flexible timetable:** Make a strategy for preparing for the CISSP exam that considers your short-term and long-term goals and any personal or professional obligations that could affect your study time. Ensure your schedule is flexible and can be changed to account for unplanned events or changes in your availability.

- **Keep track of your progress and make changes as needed.** Review your goals and deadlines often and track how close you are to reach them. If you keep falling behind or having trouble with certain areas, be ready to rethink your goals, change your timeframe, or ask for more help and resources as needed.

- **Stay realistic and prevent burnout.** It's essential to establish high goals and keep a strong commitment to preparing for the CISSP exam, but you should also be aware of the potential for fatigue. Make sure you balance work and life well by including breaks, time to relax, and self-care in your study routine.

Setting realistic goals and dates might help you build a planned and focused CISSP study strategy. This will help you make good use of your time and money. This strategy will help you stay motivated and on track as you study for the exam. It will also ensure you are well-prepared to take the CISSP exam with confidence and acquire the certification you need to advance in your cybersecurity career.

CREATING A PERSONALIZED STUDY SCHEDULE

A vital part of a good CISSP study plan is a customized study schedule. By adapting your study schedule to fit your needs, preferences, and other obligations, you may maximize your time and effort and get the most out of each study session. Here are some steps to help you create a personalized study schedule for your CISSP exam preparation:

- **Determine your available study time:** Begin by calculating the amount of time you can realistically dedicate to CISSP exam preparation each week. Consider your work, family, and social commitments, and set aside regular blocks of time for focused study sessions.

- **Establish study milestones:** Based on your short-term and long-term goals, identify key milestones for your CISSP exam preparation. These milestones could include completing specific sections of a study guide, mastering particular topics or domains, or achieving target scores on practice exams.

- **Allocate time according to priority:** When creating your study schedule, allocate more time to the topics and domains where you identified knowledge gaps or areas that require additional focus. This targeted approach will ensure that you effectively address your weaknesses and build a solid foundation across all CISSP domains.

- **Incorporate varied study activities:** To cater to your preferred learning style and maintain engagement, include a mix of study activities in your schedule, such as reading textbooks, watching video lectures, participating in discussion groups, and completing practice exams. This variety will help reinforce key concepts and keep your study sessions fresh and engaging.

- **Schedule regular breaks and downtime:** Avoid burnout by incorporating regular breaks and downtime into your study schedule. Taking short breaks during study sessions can help improve focus and retention, while setting aside time for relaxation, hobbies, and social activities will ensure that you maintain a healthy work-life balance.

- **Plan for review and revision:** Allocate time in your study schedule for periodic review and revision of the material you have covered. This will help reinforce your learning, identify any lingering knowledge gaps, and consolidate your understanding of the CISSP domains.

- **Stay flexible and adaptable:** Recognize that your study schedule may need to change as you progress through your CISSP exam preparation journey. Be prepared to adjust your schedule as needed, whether in response to personal or professional commitments or to address specific challenges that arise during your studies.

By making a personalized study program that fits your requirements and tastes, you can make the most of your time studying for the CISSP test and make sure you're ready to take it with confidence. This personalized approach will not only help you keep on track and organized as you study, but it will also set you up for a successful career in cybersecurity.

2

CISSP EXAM DOMAINS

DOMAIN 1: SECURITY AND RISK MANAGEMENT

Domain 1 of the CISSP exam, Security and Risk Management, is an essential foundation for any cybersecurity professional. This domain covers key principles, frameworks, and best practices related to the management of security and risk within an organization. Visit ISC2.org for the most up-to-date percentage weight of this domain to understand its proportion of the exam questions.. Here is a detailed breakdown of the primary topics and concepts covered within this domain:

- **Understanding and applying security concepts:** Begin by familiarizing yourself with the core security concepts, including confidentiality, integrity, and availability (CIA Triad), as well as least privilege, separation of duties, defense-in-depth, and security through obscurity. These concepts form the basis of information security and are crucial for designing and implementing effective security measures.

- **Compliance and legal considerations:** Develop an understanding of the legal, regulatory, and contractual obligations related to information security. Learn about data privacy regulations, such as GDPR and CCPA, as well as industry-specific standards like HIPAA and PCI DSS. Familiarize yourself with intellectual property laws and how they impact cybersecurity.

- **Security policies, procedures, and guidelines:** Gain a comprehensive understanding of the importance of security policies, procedures, and guidelines in maintaining an organization's security posture. Learn how to develop and implement these documents, and understand their role in supporting regulatory compliance and risk management.

- **Risk management concepts and methodologies:** Study the fundamentals of risk management, including risk identification, assessment, and mitigation strategies. Understand the role of risk appetite, risk tolerance, and risk acceptance in an organization's risk management strategy. Familiarize yourself with common risk management frameworks, such as NIST SP 800-37, ISO 31000, and FAIR.

- **Threat modeling:** Learn about various threat modeling techniques, such as STRIDE, PASTA, and VAST, and understand how they can help identify potential threats and vulnerabilities in an organization's systems and infrastructure.

- **Business continuity and disaster recovery planning:** Gain a thorough understanding of business continuity planning (BCP) and disaster recovery planning (DRP) principles. Learn how to develop, implement, and maintain these plans to minimize the impact of disruptions and ensure the continued availability of critical systems and services. Incident management: Understand the key components of an effective incident management process, including incident identification, containment, eradication, and recovery, as well as post-incident analysis and lessons learned.

- **Security awareness and training:** Recognize the importance of security awareness and training programs in reducing human-related risks and improving an organization's overall security posture. Learn about various training methodologies and best practices for developing and implementing effective security awareness programs.

By mastering the topics and concepts within Domain 1 - Security and Risk Management, you will build a strong foundation in information security principles and practices, setting the stage for success in the CISSP exam and your cybersecurity career.

DOMAIN 2: ASSET SECURITY

Domain 2 of the CISSP exam, Asset Security, focuses on the processes, best practices, and strategies required to identify, classify, and protect an organization's information assets. Visit ISC2.org for the most up-to-date percentage weight of this domain to understand its proportion of the exam questions. Here is a detailed breakdown of the primary topics and concepts covered within this domain:

- **Information and asset classification:** Learn about the importance of accurately classifying information and assets based on their value, sensitivity, and criticality. Understand the various classification levels, such as public, internal, confidential, and restricted, and the role they play in protecting sensitive information and managing access controls.

- **Asset ownership and stewardship:** Understand the concept of asset ownership and the responsibilities of asset owners, stewards, and custodians in maintaining and protecting information assets. Learn about the processes for assigning ownership and the role of data owners in ensuring appropriate security controls are in place.

- **Data protection methods:** Familiarize yourself with various data protection methods and best practices, including encryption, data masking, tokenization, and data loss prevention (DLP) technologies. Understand the advantages and disadvantages of different protection methods and how they can be applied to secure sensitive information in various scenarios.

- **Data retention, storage, and handling:** Gain a comprehensive understanding of the principles and best practices for data retention, storage, and handling. Learn about various storage media and their respective security considerations, as well as the role of data retention policies in managing the lifecycle of information assets.

- **Secure disposal and destruction:** Understand the importance of securely disposing of and destroying sensitive information and assets to prevent unauthorized access and potential data breaches. Learn about various disposal and destruction methods, including degaussing, shredding, and secure deletion, as well as the guidelines for secure disposal outlined in NIST SP 800-88.

- **Privacy protection:** Develop an understanding of privacy principles and the role of privacy protection in asset security. Familiarize yourself with privacy regulations, such as GDPR and CCPA, and their implications for data protection and privacy. Learn about privacy-enhancing technologies, including anonymization, pseudonymization, and differential privacy.

- **Data loss prevention (DLP):** Gain a thorough understanding of DLP technologies and their role in preventing unauthorized access, leakage, or misuse of sensitive information. Learn about various DLP solutions, including network-based, endpoint-based, and storage-based DLP, and how they can be deployed to protect information assets throughout their lifecycle.

By mastering the topics and concepts within Domain 2 - Asset Security, you will develop a deep understanding of the processes and strategies required to effectively protect an organization's information assets. This knowledge will not only contribute to your success in the CISSP exam but also serve as a valuable asset in your cybersecurity career.

DOMAIN 3: SECURITY ARCHITECTURE AND ENGINEERING

Domain 3 of the CISSP exam, Security Architecture and Engineering, delves into the design, implementation, and maintenance of secure systems and architectures. Visit ISC2.org for the most up-to-date percentage weight of this domain to understand its proportion of the exam questions. Here is a detailed breakdown of the primary topics and concepts covered within this domain:

- **Security models and frameworks:** Familiarize yourself with various security models and frameworks, such as the Bell-LaPadula, Biba, Clark-Wilson, and Brewer-Nash models. Understand the purpose of these models and how they help enforce confidentiality, integrity, and availability (CIA) within an organization's systems and architectures.

- **Hardware and firmware security:** Learn about the security implications of hardware and firmware components, including BIOS/UEFI, Trusted Platform Modules (TPMs), and Hardware Security Modules (HSMs). Understand the importance of secure boot processes, firmware updates, and hardware-based encryption in maintaining system security.

- **Cryptography:** Gain a comprehensive understanding of cryptographic principles, algorithms, and protocols, including symmetric and asymmetric encryption, hash functions, digital signatures, and key management. Study the applications of cryptography in securing data at rest, in transit, and in use, as well as the role of cryptography in securing network communications, such as SSL/TLS and VPNs.

- **Secure network design:** Understand the principles of secure network design, including the use of segmentation, zoning, and secure network topologies. Learn about the role of firewalls, intrusion detection and prevention systems (IDPS), and other security devices in protecting network assets.

- **Security assessment and testing:** Familiarize yourself with various security assessment and testing methodologies, such as vulnerability assessments, penetration testing, and security audits. Learn about the role of security testing in maintaining secure systems and architectures and the importance of regular assessments in identifying and mitigating vulnerabilities.

- **Security operations:** Study the principles and best practices of security operations, including patch management, change management, and incident response. Understand the role of security operations in maintaining the ongoing security and resilience of an organization's systems and infrastructure.

- **Secure software development:** Learn about secure software development principles and practices, including the Software Development Life Cycle (SDLC), secure coding practices, and software testing methodologies. Understand the importance of incorporating security into each stage of the software development process and the role of DevSecOps in ensuring software security.

- **Physical and environmental security:** Gain an understanding of physical and environmental security measures, such as access controls, surveillance systems, and environmental controls. Learn about the role of physical security in protecting an organization's systems, data, and personnel.

By mastering the topics and concepts within Domain 3 - Security Architecture and Engineering, you will develop a strong foundation in the design, implementation, and maintenance of secure systems and infrastructures. This knowledge will not only contribute to your success in the CISSP exam but also serve as a critical skillset in your cyber-security career.

DOMAIN 4: COMMUNICATION AND NETWORK SECURITY

Domain 4 of the CISSP exam, Communication and Net-work Security, focuses on the principles, technologies, and best practices for securing network communications and infrastructure. Visit ISC2.org for the most up-to-date per-centage weight of this domain to understand its proportion of the exam questions. Here is a detailed breakdown of the primary topics and concepts covered within this domain:

- **Network architecture and design principles:** Familiarize yourself with fundamental network ar-chitecture and design principles, including the OSI model, TCP/IP model, and various network to-pologies. Understand the role of these principles in designing and implementing secure network infrastructure.

- **Network components and security:** Study various network components, such as switches, routers, fire-walls, proxies, and load balancers, and understand their security implications. Learn about the impor-tance of securing network devices through proper configuration, monitoring, and patch management.

- **Secure network protocols:** Gain a comprehensive understanding of secure network protocols, includ-ing SSL/TLS, SSH, IPsec, and HTTPS. Understand the role of these protocols in securing data in transit and maintaining the confidentiality, integrity, and availability of network communications.

- **Wireless network security:** Learn about the unique security challenges associated with wireless networks, such as rogue access points, war driving, and evil twin attacks. Understand the importance of implementing strong encryption and authentication mechanisms, such as WPA3 and 802.1X, to secure wireless communications.

- **Network segmentation and isolation:** Study the principles of network segmentation and isolation, including the use of Virtual Local Area Networks (VLANs), subnets, and demilitarized zones (DMZs). Understand the role of segmentation in limiting the potential impact of security incidents and improving overall network security.

- **Intrusion detection and prevention:** Familiarize yourself with intrusion detection and prevention systems (IDPS), including signature-based, anomaly-based, and behavior-based detection methods. Learn about the importance of deploying IDPS solutions to monitor and protect network infrastructure from potential threats.

- **Remote access and Virtual Private Networks (VPNs):** Gain an understanding of remote access technologies and their security implications, including Remote Desktop Protocol (RDP), Virtual Network Computing (VNC), and Secure Shell (SSH).

Learn about the role of VPNs in providing secure, encrypted connections over untrusted networks, and study various VPN technologies, such as IPsec and SSL/TLS-based VPNs.

- **Network security monitoring and analysis:** Understand the importance of monitoring and analyzing network traffic to detect potential security incidents and maintain overall network security. Learn about various network monitoring and analysis tools, such as network analyzers, packet capture utilities, and Security Information and Event Management (SIEM) systems.

By mastering the topics and concepts within Domain 4 - Communication and Network Security, you will develop a deep understanding of the technologies and best practices required to secure network communications and infrastructure. This knowledge will not only contribute to your success in the CISSP exam but also serve as a valuable asset in your cybersecurity career.

DOMAIN 5: IDENTITY AND ACCESS MANAGEMENT (IAM)

Domain 5 of the CISSP exam, Identity and Access Management (IAM), focuses on the processes, technologies, and best practices for managing access to an organization's resources and systems. Visit ISC2.org for the most up-to-date percentage weight of this domain to understand its proportion of the exam questions. Here is a detailed breakdown of the primary topics and concepts covered within this domain:

- **Identity management concepts:** Familiarize yourself with the fundamentals of identity management, including the concepts of identification, authentication, authorization, and accountability (IAAA). Understand the differences between these concepts and their roles in maintaining the security of an organization's resources.

- **Access control models:** Study various access control models, such as Discretionary Access Control (DAC), Mandatory Access Control (MAC), Role-Based Access Control (RBAC), and Attribute-Based Access Control (ABAC). Learn how these models can be applied to manage access to resources and systems based on different requirements and constraints.

- **Authentication factors and methods:** Gain a comprehensive understanding of different authentication factors (something you know, something you have, and something you are) and methods, including passwords, tokens, biometrics, and multi-factor authentication (MFA). Understand the strengths and weaknesses of different authentication methods and how they can be combined to enhance security.

- **Access control technologies:** Familiarize yourself with various access control technologies, such as RADIUS, TACACS+, Kerberos, LDAP, and OAuth. Learn how these technologies work, their security implications, and their roles in managing access to resources and systems.

- **Identity and access provisioning:** Understand the importance of identity and access provisioning, including the processes of user registration, onboarding, offboarding, and access modification. Learn about the role of Identity and Access Management (IAM) systems in automating and streamlining these processes and reducing the risk of unauthorized access.

- **Access control monitoring and auditing:** Study the principles and best practices for monitoring and auditing access control, including the use of log management, Security Information and Event Management (SIEM) systems, and access control audits. Understand the importance of regular monitoring and auditing in detecting and mitigating unauthorized access and potential security incidents.

- **Federation and Single Sign-On (SSO):** Learn about the concepts of federation and Single Sign-On (SSO), including the benefits of using federated identity and SSO solutions, such as reduced authentication overhead and improved user experience. Familiarize yourself with common federated identity and SSO technologies, such as SAML, OpenID Connect, and Microsoft Active Directory Federation Services (AD FS).

By mastering the topics and concepts within Domain 5 - Identity and Access Management (IAM), you will develop a strong foundation in the processes and technologies required to manage access to an organization's resources and systems. This knowledge will not only contribute to your success in the CISSP exam but also serve as a critical skillset in your cybersecurity career.

DOMAIN 6: SECURITY ASSESSMENT AND TESTING

Domain 6 of the CISSP exam, Security Assessment and Testing, focuses on the methodologies, techniques, and tools used to evaluate the effectiveness of an organization's security controls and identify potential vulnerabilities. Visit ISC2.org for the most up-to-date percentage weight of this domain to understand its proportion of the exam questions. Here is a detailed breakdown of the primary topics and concepts covered within this domain:

- **Security assessment methodologies:** Familiarize yourself with various security assessment methodologies, such as vulnerability assessments, penetration testing, and security audits. Understand the objectives, scope, and limitations of each methodology, and learn how they can be applied to evaluate an organization's security posture.

- **Vulnerability assessment tools and techniques:** Gain a comprehensive understanding of different vulnerability assessment tools and techniques, including automated vulnerability scanners, manual testing, and configuration reviews. Learn about the benefits and limitations of different tools and techniques and how they can be combined to provide a more comprehensive assessment of an organization's security.

- **Penetration testing concepts and methodologies:** Study the concepts and methodologies of penetration testing, including black-box, white-box, and gray-box testing, as well as various testing techniques, such as social engineering, network, and application testing. Understand the role of penetration testing in identifying potential vulnerabilities and validating the effectiveness of security controls.

- **Security audits and reviews:** Learn about the process of conducting security audits and reviews, including the objectives, scope, and methods used to evaluate an organization's security policies, procedures, and controls. Understand the role of internal and external auditors in maintaining compliance with regulatory and industry standards, such as ISO 27001, PCI DSS, and HIPAA.

- **Log reviews and SIEM:** Familiarize yourself with the importance of log reviews and Security Information and Event Management (SIEM) systems in monitoring, detecting, and responding to potential security incidents. Learn about the features and functions of SIEM solutions, as well as best practices for log management and analysis.

- **Incident response testing and exercises:** Understand the importance of testing and exercising incident response plans to validate their effectiveness and identify potential areas for improvement. Learn about various incident response testing methods, such as tabletop exercises, simulations, and full-scale exercises, and how they can be used to enhance an organization's incident response capabilities.

- **Security control testing and validation:** Gain a thorough understanding of the process of testing and validating security controls, including functional and non-functional testing methods, such as performance, usability, and resilience testing. Learn about the role of testing and validation in ensuring the effectiveness and reliability of security controls and mitigating potential risks.

By mastering the topics and concepts within Domain 6 - Security Assessment and Testing, you will develop a deep understanding of the methodologies, techniques, and tools used to evaluate an organization's security posture and identify potential vulnerabilities. This knowledge will not only contribute to your success in the CISSP exam but also serve as a valuable asset in your cybersecurity career.

DOMAIN 7: SECURITY OPERATIONS

Domain 7 of the CISSP exam, Security Operations, covers the principles, processes, and best practices for maintaining the security, availability, and integrity of an organization's information systems and infrastructure. Visit ISC2.org for the most up-to-date percentage weight of this domain to understand its proportion of the exam questions. Here is a detailed breakdown of the primary topics and concepts covered within this domain:

- **Incident response and management:** Familiarize yourself with the principles and processes of incident response and management, including the steps of incident identification, containment, eradication, recovery, and lessons learned. Understand the role of incident response teams, communication strategies, and the importance of having a well-defined incident response plan.

- **Business continuity and disaster recovery planning:** Learn about the concepts of business continuity and disaster recovery planning, including the steps of risk assessment, business impact analysis (BIA), and the development of recovery strategies. Understand the differences between Business Continuity Plans (BCPs) and Disaster Recovery Plans (DRPs) and the importance of testing and maintaining these plans.

- **Change management:** Study the principles and best practices of change management, including the steps of change request, approval, implementation, and review. Understand the role of change management in maintaining system stability and security, as well as the importance of documenting and communicating changes.

- **Security operations centers (SOCs):** Gain a comprehensive understanding of the functions and responsibilities of a Security Operations Center (SOC), including the roles of analysts, engineers, and managers. Learn about the benefits of having a dedicated SOC and the technologies used to monitor and respond to potential security incidents.

- **Secure asset disposal:** Familiarize yourself with the importance of secure asset disposal, including the proper disposal of hardware, software, and data.

Understand the role of asset disposal in mitigating the risks of data leakage and ensuring compliance with data protection regulations.

- **Patch and vulnerability management:** Learn about the processes of patch and vulnerability management, including the identification, assessment, and remediation of vulnerabilities. Understand the importance of timely patching and the challenges of managing patches in complex environments.

- **Log management and monitoring:** Study the principles and best practices for log management and monitoring, including the collection, storage, analysis, and reporting of log data. Understand the role of logs in detecting and responding to security incidents, as well as the importance of protecting log data from unauthorized access and tampering.

- **Data backup and recovery:** Gain an understanding of data backup and recovery strategies, including the selection of appropriate backup media, schedules, and recovery methods. Learn about the importance of regular backups in ensuring data availability and integrity, as well as the challenges of managing backups in large-scale environments.

By mastering the topics and concepts within Domain 7 - Security Operations, you will develop a strong foundation in the principles, processes, and best practices required to maintain the security, availability, and integrity of an organization's information systems and infrastructure. This knowledge will not only contribute to your success in the CISSP exam but also serve as a critical skillset in your cybersecurity career.

DOMAIN 8: SOFTWARE DEVELOPMENT SECURITY

Domain 8 of the CISSP exam, Software Development Security, focuses on the principles, best practices, and methodologies for integrating security into software development processes. Visit ISC2.org for the most up-to-date percentage weight of this domain to understand its proportion of the exam questions. Here is a detailed breakdown of the primary topics and concepts covered within this domain:

- **Secure software development methodologies:** Familiarize yourself with various secure software development methodologies, such as the Waterfall model, Agile, DevOps, and DevSecOps. Understand the role of these methodologies in ensuring the development of secure software and reducing the risks associated with software vulnerabilities.

- **Software development lifecycle (SDLC):** Study the stages of the Software Development Lifecycle (SDLC), including requirements analysis, design, implementation, testing, and maintenance. Understand the importance of integrating security considerations into each stage of the SDLC to minimize potential vulnerabilities and risks.

- **Security requirements and design principles:** Learn about the process of identifying and documenting security requirements, including functional and non-functional requirements, as well as best practices for designing secure software, such as the principles of least privilege, defense-in-depth, and separation of duties.

- **Secure coding practices and guidelines:** Gain a comprehensive understanding of secure coding practices and guidelines, such as input validation, output encoding, error handling, and secure session management. Familiarize yourself with common coding vulnerabilities, such as SQL injection, cross-site scripting (XSS), and buffer overflows, and how to prevent them.

- **Application security testing and validation:** Study the principles and techniques of application security testing and validation, including static and dynamic analysis, fuzz testing, and penetration testing. Understand the role of security testing in identifying potential vulnerabilities and validating the effectiveness of security controls.

- **Software vulnerability management:** Learn about the process of managing software vulnerabilities, including the identification, assessment, and remediation of vulnerabilities in software components, dependencies, and third-party libraries. Understand the importance of timely patching and the challenges of managing vulnerabilities in complex software environments.

- **Application security controls and countermeasures:** Familiarize yourself with various application security controls and countermeasures, such as access control, encryption, and secure storage. Learn about the role of these controls in mitigating potential risks and ensuring the confidentiality, integrity, and availability of application data.

- **Secure software deployment and operations:** Gain an understanding of the principles and best practices for secure software deployment and operations, including the use of secure configuration management, continuous integration and continuous delivery (CI/CD) pipelines, and security monitoring and incident response.

By mastering the topics and concepts within Domain 8 - Software Development Security, you will develop a deep understanding of the principles, best practices, and methodologies for integrating security into software development processes. This knowledge will not only contribute to your success in the CISSP exam but also serve as a valuable asset in your cybersecurity career.

STUDY MATERIALS AND RESOURCES

THE OFFICIAL (ISC)^2 CISSP STUDY GUIDE

The Official (ISC)^2 CISSP Study Guide is a thorough and reliable resource that covers all eight of the CISSP exam's topics in depth. The (ISC)^2 organization, which is in charge of the CISSP certification, produced the study guide to help candidates learn the skills and information they need to pass the exam and get the certification. In this section, we will explore what the study guide offers and how to use it effectively as a study material and resource for CISSP preparation.

- What is the Official (ISC)^2 CISSP Study Guide?
  - **The Official (ISC)^2 CISSP Study Guide** is a detailed textbook that covers each of the eight domains of the CISSP exam, providing in-depth explanations of the concepts, principles, and best practices within each domain. The study guide is written by certified CISSP professionals and is updated regularly to reflect the latest exam objectives and industry developments.

- How to use the Official (ISC)^2 CISSP Study Guide as a study material and resource:
  - **Read the study guide thoroughly:** Begin by reading the entire study guide from cover to cover, focusing on understanding the concepts, principles, and best practices within each domain. Take notes on key points and highlight any areas that require further study or clarification.
  - **Test your knowledge with practice questions:** The Official (ISC)^2 CISSP Study Guide includes practice questions at the end of each chapter, which can help you assess your understanding of

the material and identify any areas that require additional study. Answer these questions and review the explanations provided for both correct and incorrect answers to reinforce your understanding of the concepts.

- **Review the study guide periodically:** As you progress through your CISSP preparation, periodically review the study guide to refresh your memory and ensure that you have a solid understanding of all the concepts, principles, and best practices within each domain.
- **Use the study guide as a reference:** The Official (ISC)² CISSP Study Guide can also serve as a valuable reference resource throughout your cybersecurity career. Keep the study guide handy to consult whenever you encounter unfamiliar concepts or require a refresher on specific topics.
- **Supplement with additional resources:** While the Official (ISC)² CISSP Study Guide is an excellent primary resource for CISSP preparation, it is important to supplement your studies with additional resources, such as online forums, video courses, and practice exams, to ensure a well-rounded understanding of the exam material.

By using the Official (ISC)² CISSP Study Guide as a crucial part of your CISSP preparation plan, you will understand the ideas, principles, and best practices you need to know to pass the test and get the CISSP certification. Also, the study guide will be an excellent resource for you throughout your cybersecurity career, helping you keep up with the latest news and best practices.

THE CISSP ALL-IN-ONE EXAM GUIDE

The CISSP All-in-One Exam Guide is a popular and well-known way to prepare for the CISSP exam. The guide, written by Shon Harris and updated by Fernando Maymi, gives a complete and helpful way to study for the CISSP exam. In this section, we'll discuss what the CISSP All-in-One Exam Guide offers and how to use it as a study tool and resource to pass the CISSP exam.

- What is the CISSP All-in-One Exam Guide?
  - **The CISSP All-in-One Exam Guide:** is a detailed and comprehensive textbook covering all eight domains of the CISSP exam. The guide is known for its clear and concise explanations, practical examples, and real-world insights, making it an invaluable resource for CISSP candidates. The guide is updated regularly to ensure that it aligns with the latest exam objectives and industry developments.

- How to use the CISSP All-in-One Exam Guide as a study material and resource:
  - **Read the guide thoroughly:** Start by reading the entire CISSP All-in-One Exam Guide from beginning to end, focusing on understanding the concepts, principles, and best practices within each domain. Take notes on key points and highlight any areas that require further study or clarification.
  - **Test your knowledge with practice questions:** The CISSP All-in-One Exam Guide includes practice questions at the end of each chapter, which can help you assess your understanding of the

material and identify any areas that require additional study. Answer these questions and review the explanations provided for both correct and incorrect answers to reinforce your understanding of the concepts.

- **Use the guide's practical examples:** The CISSP All-in-One Exam Guide is known for its practical examples and real-world insights, which can help you better understand the application of CISSP concepts in real-life scenarios. Pay close attention to these examples as you read through the guide, and try to relate them to your own experiences and professional context.

- **Review the guide periodically:** As you progress through your CISSP preparation, periodically review the CISSP All-in-One Exam Guide to refresh your memory and ensure that you have a solid understanding of all the concepts, principles, and best practices within each domain.

- **Supplement with additional resources:** While the CISSP All-in-One Exam Guide is an excellent primary resource for CISSP preparation, it is essential to supplement your studies with additional resources, such as the Official (ISC)[2] CISSP Study Guide, online forums, video courses, and practice exams, to ensure a well-rounded understanding of the exam material.

Using the CISSP All-in-One Exam Guide as part of your CISSP study plan will teach you the ideas, principles, and best practices you need to know to pass the exam and get the CISSP certification. The guide will also be a valuable resource for your cybersecurity career. It will help you stay up-to-date on the sector's newest innovations and best practices.

QUALITY CISSP PRACTICE EXAMS

CISSP Practice Exams are essential to getting ready for the CISSP certification exam. They allow applicants to see how much they know, figure out what they need to learn more about and gain practice answering questions like those on an exam. In this section, we'll talk about a few places to get good CISSP Practice Exams and how to use them as an excellent way to study and learn for the CISSP exam.

- **Official (ISC)² CISSP Practice Tests:** The (ISC)² organization, responsible for the CISSP certification, offers official CISSP Practice Tests. These practice tests include questions that align closely with the actual exam objectives and are written by (ISC)²-certified professionals. Using the official practice tests can give you a good understanding of the question formats and the level of difficulty you can expect on the actual exam.

- **CISSP All-in-One Exam Guide:** As mentioned previously, the CISSP All-in-One Exam Guide includes practice questions at the end of each chapter. These questions, along with their detailed explanations, can help reinforce your understanding of the exam material and provide additional practice.

- **Boson ExSim-Max for CISSP:** Boson's ExSim-Max for CISSP is a highly regarded practice exam resource that offers a comprehensive set of practice questions, simulations, and detailed explanations. The practice exams are designed to mimic the actual CISSP exam format and difficulty level, making them a valuable resource in gauging your readiness for the test.

- **CCCure.org:** CCCure is a popular platform that provides a wide range of CISSP practice questions and quizzes, along with explanations and references. The platform allows you to customize your quizzes based on your preferred domains and difficulty levels, making it a versatile study tool.

- **uCertify CISSP Practice Questions:** uCertify offers a comprehensive set of CISSP practice questions designed to closely resemble the actual exam format and content. These questions cover all eight domains of the CISSP Common Body of Knowledge (CBK) and are regularly updated to ensure their relevance and accuracy. The practice questions are developed by certified professionals and industry experts, ensuring that they provide a true representation of the challenges candidates will face on the actual exam.

How to use CISSP Practice Exams effectively as a study material and resource:

- **Take practice exams regularly:** Incorporate practice exams into your study schedule, taking them regularly to assess your progress and identify areas that need improvement. This will also help you familiarize yourself with the exam format and time constraints.

- **Review the explanations:** After completing a practice exam, carefully review the explanations for both correct and incorrect answers. This will help you understand the reasoning behind the answers and reinforce your knowledge of the exam material.

- **Learn from your mistakes:** Identify any recurring patterns or specific areas where you struggle, and focus on improving your understanding of those concepts. Use your practice exam results to guide your study efforts and allocate more time to areas that require additional attention.

- **Track your progress:** Monitor your practice exam scores over time to gauge your progress and determine whether you are ready to take the actual CISSP exam. Remember that practice exams are just one aspect of your CISSP preparation and should be supplemented with other study materials and resources.

By utilizing quality CISSP Practice Exams from various sources as part of your CISSP preparation strategy, you will gain valuable experience in answering exam-style questions, identify areas that require further study, and build your confidence for the actual CISSP exam.

QUALITY ONLINE VIDEO COURSES AND TUTORIALS FOR CISSP PREPARATION

Online video courses and tutorials are becoming increasingly popular as ways to study for the CISSP exam. They offer a dynamic and exciting way to learn, which helps applicants understand complex ideas and principles. In this section, we'll talk about a few places where you may find good online video courses and tutorials for CISSP preparation, as well as how to make the most of them as a way to study and learn.

- **(ISC)² Official CISSP Online Self-Paced Training:** The official (ISC)² CISSP online self-paced training offers a comprehensive and interactive learning experience, covering all eight domains of the CISSP exam. The course includes video lectures, interactive flashcards, quizzes, and other engaging learning materials, all delivered by (ISC)²-certified instructors.

- **Cybrary CISSP Training:** Cybrary offers a free CISSP video course designed to provide a thorough understanding of the CISSP exam domains. The course includes video lectures, supplemental readings, and quizzes. Cybrary also offers a paid membership plan that includes access to additional resources such as practice exams and virtual labs.

- **Pluralsight CISSP Training:** Pluralsight provides a comprehensive CISSP training course consisting of video lectures, quizzes, and practical exercises. The course is designed to cover each of the eight CISSP exam domains in detail, with a strong emphasis on real-world examples and practical application.

- **LinkedIn Learning (formerly Lynda.com):** LinkedIn Learning offers a CISSP preparation course that includes in-depth video lectures, quizzes, and additional resources for each of the eight CISSP exam domains. The course is designed for professionals seeking a comprehensive understanding of the CISSP exam material and is taught by experienced CISSP-certified instructors.

How to use online video courses and tutorials effectively as a study material and resource:

- **Supplement your reading materials:** Online video courses and tutorials can be used to complement your primary study materials, such as the Official (ISC)² CISSP Study Guide or the CISSP All-in-One Exam Guide. They offer a different perspective and learning approach, which can help reinforce your understanding of the exam material.

- **Make a study schedule:** Allocate specific time slots for watching video lectures and tutorials as part of your study schedule. Consistency is key, and setting aside regular time for video-based learning can help you stay on track and maintain your momentum throughout your CISSP preparation.

- **Take notes and review:** While watching video lectures and tutorials, take notes on key concepts, principles, and best practices. Review these notes regularly to reinforce your understanding and retention of the exam material.

- **Participate in online discussions:** Many online video courses and tutorials offer discussion forums or platforms where you can engage with other CISSP candidates and instructors. Participate in these discussions to ask questions, share insights, and learn from the experiences of others.

Use good online video courses and tutorials to prepare for the CISSP exam. You will have a dynamic and exciting learning experience that can help you understand the CISSP subject better. Using these tools, along with other study materials and practice exams, can help you prepare for the CISSP exam in a well-rounded and thorough way.

CISSP STUDY GROUPS AND FORUMS - A VALUABLE SUPPORT SYSTEM FOR CISSP PREPARATION

People studying for the CISSP exam can get a lot out of CISSP study groups and forums. You can share your experiences, ask questions, and learn from people working for CISSP certification on these platforms. In this part, we'll talk about some of the most popular CISSP study groups and forums and how to make the most of them when studying for the CISSP exam.

- **(ISC)² Official CISSP Community:** The (ISC)² organization maintains an official CISSP Community forum where candidates can engage in discussions, ask questions, and share their experiences. The forum is moderated by (ISC)²-certified professionals, ensuring that the information and advice shared is accurate and up-to-date.

- **TechExams CISSP Forum:** TechExams is a popular platform for IT certification discussions, including a dedicated CISSP forum. The forum is frequented by CISSP candidates, certified professionals, and industry experts, providing a wealth of knowledge and experience to draw upon.

- **Reddit CISSP Community:** The CISSP subreddit (/r/cissp) is an active community of CISSP candidates, certified professionals, and industry experts. This platform offers a wealth of information, advice, and support for those preparing for the CISSP exam.

- **CISSP Study Groups on Social Media:** Many CISSP study groups can be found on social media platforms like Facebook and LinkedIn. These groups offer a more informal setting where you can connect with other CISSP candidates, share study tips, and discuss exam-related topics.

How to use CISSP study groups and forums effectively as a study material and resource:

- **Ask questions:** Don't hesitate to ask questions in study groups and forums. This is an excellent way to clarify your understanding of complex concepts and learn from the experiences of others who have faced similar challenges.

- **Share your knowledge:** As you progress through your CISSP preparation, share your knowledge and insights with others in the study groups and forums. By doing so, you not only help others but also reinforce your own understanding of the exam material.

- **Stay engaged:** Regularly participate in discussions and stay engaged with the study group or forum community. This will help you stay motivated and maintain momentum throughout your CISSP preparation journey.

- **Learn from others' experiences:** Take the time to read about the experiences of others who have taken the CISSP exam, as this can provide valuable insights into what to expect on exam day and how to approach your preparation.

By joining CISSP study groups and forums, you may get support from your peers and specialists who can help you figure out how to study for the CISSP exam. These platforms significantly add to existing study materials, practice exams, and online video courses. Together, they help you prepare for the CISSP exam in a complete and well-rounded way.

MOBILE APPS AND FLASHCARDS - CONVENIENT AND EFFECTIVE CISSP STUDY TOOLS

Mobile apps and flashcards are easy and effective ways to study for the CISSP exam. They give applicants a flexible and easy way to check on the go. In this part, we'll talk about some popular flashcards and mobile apps that can help you study for the CISSP exam and how to use them efficiently.

- **Official (ISC)² CISSP Flashcards:** The (ISC)² organization offers official CISSP Flashcards, available as a mobile app for iOS and Android devices. The flash-cards cover all eight domains of the CISSP exam and are designed to help you reinforce your understand-ing of key concepts, principles, and best practices.

- **Quizlet CISSP Flashcards:** Quizlet is a popular plat-form for user-generated flashcards, including many CISSP-related sets created by candidates and cer-tified professionals. You can search for CISSP flash-cards on Quizlet and use their mobile app to study anytime, anywhere.

- **CISSP Pocket Prep:** The CISSP Pocket Prep app is available for iOS and Android devices and offers a range of features, including customizable practice exams, flashcards, and performance tracking. The app includes hundreds of practice questions de-signed to test your knowledge and help you identify areas that require further study.

- **CISSP Practice Test & Flashcards:** This mobile app, available for iOS and Android devices, combines practice test questions with flashcards to create a comprehensive CISSP study tool. The app includes over 1,000 practice questions and flashcards, cover-ing all eight CISSP exam domains.

How to use mobile apps and flashcards effectively as a study material and resource:

- **Study in short bursts:** Mobile apps and flashcards are designed for quick and convenient study ses-sions, allowing you to make the most of your spare time. Use these resources during your daily commute, lunch breaks, or any other time you have a few minutes to spare.

- **Focus on specific domains:** Many mobile apps and flashcards allow you to filter content based on specific CISSP exam domains. Use this feature to focus on areas where you need extra practice or reinforcement.

- **Test your knowledge regularly:** Use mobile apps and flashcards to quiz yourself regularly and gauge your progress. This will help you identify areas that require additional study and keep your memory fresh.

- **Combine with other study materials:** Mobile apps and flashcards are excellent supplementary resources but should not be used in isolation. Combine them with other study materials, such as textbooks, online video courses, practice exams, and study groups, to ensure a well-rounded and comprehensive approach to your CISSP exam preparation.

Using mobile apps and flashcards as part of your CISSP preparation plan will give you a flexible and easy-to-use way to study that can help you remember essential ideas, principles, and best practices. When utilized with other study materials and resources, these tools will help you prepare for the CISSP exam in a well-rounded and complete way.

NIST PUBLICATIONS AND INDUSTRY RESOURCES - ESSENTIAL CISSP STUDY MATERIAL AND RESOURCES

The CISSP exam examines candidates on what they know about the eight domains and how well they grasp security norms and practices in the actual world. For a profound grasp of the security landscape and to prepare for the CISSP exam, you need to study NIST publications and other industry resources. In this section, we'll talk about a few NIST documents and help from the industry, as well as how to use them to study for the CISSP exam.

- **NIST Special Publications (SPs):** NIST SPs are a series of documents that provide guidelines, rec-ommendations, and technical specifications relat-ed to information security. Key NIST SPs relevant to the CISSP exam include SP 800-53 (Security and Privacy Controls for Federal Information Systems and Organizations), SP 800-37 (Risk Management Framework for Information Systems and Organizations), and SP 800-30 (Guide for Conducting Risk Assessments). Reviewing these publications will provide you with a solid under-standing of industry best practices and standards.

- **ISO/IEC Standards:** The International Organization for Standardization (ISO) and the International Electrotechnical Commission (IEC) have published numerous standards related to information securi-ty. Key standards relevant to the CISSP exam include ISO/IEC 27001 (Information Security Management Systems) and ISO/IEC 27002 (Code of Practice for Information Security Controls). Familiarize yourself with these standards to understand the global best practices for information security management.

- **SANS Institute Resources:** The SANS Institute is a leading provider of information security training, research, and certification programs. They offer numerous resources, including white papers, webcasts, and research papers, which can be valuable study materials for CISSP candidates. Visit their website to explore their extensive library of information security resources.

- **Center for Internet Security (CIS) Critical Security Controls:** The CIS Critical Security Controls are a set of prioritized actions designed to improve an organization's security posture. Understanding these controls and their practical application can be beneficial when preparing for the CISSP exam.

How to use NIST publications and industry resources effectively as a study material and resource:

- **Focus on key documents:** While there are numerous NIST publications and industry resources available, focus on the most relevant documents for the CISSP exam, as mentioned above. Start by familiarizing yourself with the key concepts and recommendations, and then delve deeper into the specifics as needed.

- **Understand real-world applications:** As you review NIST publications and industry resources, focus on understanding how the guidelines and recommendations are applied in real-world scenarios. This will not only help you with the CISSP exam but also enhance your overall understanding of information security.

- **Supplement with other study materials:** NIST publications and industry resources should be used alongside other study materials, such as textbooks, online video courses, practice exams, study groups, and flashcards. Combining these resources will ensure a comprehensive and well-rounded approach to your CISSP exam preparation.

- **Take notes and summarize:** While reviewing NIST publications and industry resources, take notes on key concepts, principles, and best practices. Summarizing these documents in your own words can help reinforce your understanding and retention of the material.

By studying for the CISSP exam using NIST publications and industry resources, you will learn more about best practices and standards for information security. When utilized with other study tools, these resources will help you prepare for the CISSP exam in a well-rounded and thorough way.

STUDY STRATEGIES AND TECHNIQUES

ACTIVE LEARNING AND NOTE-TAKING STRATEGIES AND TECHNIQUES FOR CISSP EXAM PREPARATION

Active studying and taking good notes are essential for remembering information and ensuring you comprehend everything on the CISSP exam. In this section, we'll talk about different strategies and techniques for active learning and taking notes that can help you get the most out of your studying and improve your overall CISSP exam preparation.

- **Active Reading:** To ensure that you are actively engaging with the material, read with a purpose. Ask yourself questions before, during, and after reading a section. Highlight key points and write down any questions or thoughts that come to mind. This will help you retain the information and encourage critical thinking.

- **Cornell Method of Note-taking:** The Cornell Method involves dividing your note paper into two columns – the left-hand column for main ideas or questions, and the right-hand column for supporting details or answers. After completing your notes, summarize the main points at the bottom of the page. This technique helps to organize and consolidate information, making it easier to review and retain.

- **Mind Mapping:** Mind mapping is a visual note-taking technique that helps you organize information in a hierarchical structure, connecting related concepts and ideas. Start with a central topic, and then create branches for subtopics, further branching out for supporting details. This method can help you identify relationships between different concepts and improve your understanding of the CISSP exam material.

- **Practice Teaching:** One of the most effective ways to solidify your understanding of CISSP material is to practice teaching it to someone else, or even to yourself. This forces you to organize and articulate the information, reinforcing your comprehension and retention of the material.

- **Self-Quizzing:** Regularly quiz yourself on the material you have learned. Create flashcards, use practice questions from study guides or mobile apps, or ask yourself questions about the material. This not only tests your knowledge but also improves your long-term memory retention.

- **Study Groups:** Participating in study groups allows you to discuss CISSP exam material with others, ask questions, and share your knowledge. Engaging in active discussions with peers can help reinforce your understanding of the material and expose you to different perspectives and interpretations.

- **Regular Review:** Schedule regular reviews of your notes and study materials. This will help reinforce your memory of the material and identify any areas that require further study.

To get the most out of these tactics and approaches for active learning and taking notes, mix and match them as needed to fit your learning style and preferences. By doing things with the CISSP test content, you will learn it better and remember it longer. This will help you prepare for the CISSP exam in a well-rounded and complete way.

Understanding Key Concepts and Terminology in CISSP Exam Preparation

To do well on the CISSP exam, you need to understand important ideas and terms. In this section, we'll discuss different strategies and techniques to help you comprehend and remember the essential concepts and phrases you need to know for the CISSP exam.

- **Create a Glossary:** As you study, create a glossary of essential CISSP terms and concepts, along with their definitions and examples. This will not only help you track your progress in learning new terms but will also serve as a valuable reference for future review.

- **Break Down Complex Concepts:** When faced with a complicated concept or idea, break it down into smaller, manageable parts. Focus on understanding each part before moving on to the next. This will make it easier to grasp the overall concept and help you retain the information more effectively.

- **Use Analogies and Real-World Examples:** Relating abstract concepts to real-world examples or analogies can make them more accessible and easier to understand. Look for examples in your study materials, or think of your own situations that illustrate the concept in a practical context.

- **Visualize Concepts:** Visual aids, such as diagrams, flowcharts, and mind maps, can help clarify complex concepts and relationships between different ideas. Use these tools to represent the information visually and to reinforce your understanding.

- **Leverage Online Resources:** Utilize online resources, such as articles, blogs, and YouTube videos, to find alternative explanations or demonstrations of complex concepts and terminology. Sometimes, hearing or seeing information presented in a different way can help solidify your understanding.

- **Practice Applying Concepts:** Apply the concepts you've learned to real-world scenarios or hypothetical situations. This not only tests your understanding but also helps to reinforce the knowledge by putting it into practice.

- **Review and Test Frequently:** Regularly review the key concepts and terminology you've learned and test yourself on them. This can be done through self-quizzing, flashcards, or practice exams. Frequent review and testing will help reinforce your memory and understanding of the material.

- **Seek Clarification:** If you're struggling to understand a particular concept or term, don't hesitate to ask for help. Reach out to study group members, online forums, or instructors to seek clarification and deepen your understanding.

By using these tips and tricks, you'll be able to learn and remember the main ideas and terms you need to know for the CISSP exam. Remember that you need to know these basic things to do well on the exam and in your job as an information security expert. Combining these tactics with other study methods and resources can help you prepare for the CISSP exam in a well-rounded and thorough way.

APPLYING REAL-WORLD SCENARIOS AND CASE STUDIES IN CISSP EXAM PREPARATION

The CISSP exam is meant to assess how much you know about information security ideas and practices and how well you can use what you know in the real world. You may better understand the topic and prepare for the test by studying real-world scenarios and case studies. This section will discuss different ways to look for the CISSP exam that uses real-world scenarios and case studies.

- **Review Case Studies:** Start by reviewing case studies and real-world examples found in your study materials, such as textbooks, online courses, and industry resources. These case studies can help illustrate how the concepts you're learning are applied in actual situations, making the material more relatable and easier to understand.

- **Analyze and Discuss Scenarios:** Take the time to analyze and discuss real-world scenarios with your study group or peers. Examine the situation, identify the security issues involved, and discuss possible solutions based on the CISSP exam material. This active engagement will help reinforce your understanding of the material and develop your problem-solving skills.

- **Create Your Own Scenarios:** Develop hypothetical scenarios based on the CISSP exam domains and think through how you would apply the concepts you've learned to address the issues presented. This exercise helps you practice applying the material in a real-world context and can reveal any gaps in your understanding.

- **Learn from Real-World Incidents:** Keep up to date with current events and news related to information security. Study high-profile incidents, data breaches, and security threats to understand how the CISSP exam concepts apply to real-world situations. Analyze the causes and consequences of these incidents, as well as the measures taken to prevent or remediate the issues.

- **Relate to Personal Experiences:** Reflect on your own professional experiences, and consider how the CISSP exam material relates to your past or current work. This will help make the material more relevant and meaningful, and can also provide additional insights into how the concepts apply in practice.

- **Practice Exam Scenarios:** Many CISSP practice exams include scenario-based questions, which require you to apply your knowledge to real-world situations. Take advantage of these questions to practice your problem-solving skills and gain experience in applying CISSP concepts to practical scenarios.

By preparing for the CISSP exam with real-world scenarios and case studies, you will gain a better knowledge of the content and be better able to use the ideas in the real world. When used with additional study methods and materials, these ideas and techniques will help you prepare for the CISSP exam in a well-rounded and complete way.

Regularly Reviewing and Reinforcing Material in CISSP Exam Preparation

Reviewing and reinforcing the CISSP exam material regularly is essential for remembering the information and ensuring you fully grasp it. In this section, we will talk about many ways to properly study and reinforce the CISSP information while you prepare for the exam.

- **Spaced Repetition:** Spaced repetition is a learning technique that involves reviewing material at increasing intervals over time. By spacing out your reviews, you can improve your long-term retention of the material. Create a review schedule based on the spaced repetition principle, incorporating regular reviews of each domain and key concepts.

- **Active Recall:** Active recall involves retrieving information from your memory rather than passively reviewing it. This can be done through self-quizzing, flashcards, or practice exams. By actively recalling the material, you strengthen the neural pathways in your brain, making it easier to remember the information in the future.

- **Summary Sheets:** Create summary sheets or cheat sheets for each CISSP domain, including key concepts, terminology, and best practices. Regularly review these summaries to reinforce your understanding of the material and to identify any areas that require further study.

- **Teach Others:** Teaching the material to others or explaining it in your own words can help reinforce your understanding and improve your retention of the material. Consider discussing key concepts with a study group, peer, or even recording yourself explaining the material.

- **Review with Variety:** Mix up your review methods to keep things interesting and maintain your motivation. Alternate between self-quizzing, discussing concepts with a study group, watching video tutorials, or working through practice exams. This variety can help you avoid burnout and maintain your focus on the material.

- **Progress Tracking:** Keep track of your progress as you review the CISSP exam material. This can help you identify areas where you need to focus your efforts and can also provide a sense of accomplishment as you see your knowledge and understanding grow.

- **Set Review Milestones:** Set review milestones throughout your study plan, such as completing a certain number of practice exams or fully understanding a specific domain. These milestones can help you maintain your focus and provide motivation to continue reviewing and reinforcing the material.

Using these strategies and methods to prepare for the CISSP exam, you may efficiently review and reinforce the subject, ensuring you fully understand it. Regular review and reinforcement, together with other study methods and resources, will help you prepare for the CISSP exam in a well-rounded and successful way.

UTILIZING PRACTICE EXAMS AND SIMULATORS IN CISSP EXAM PREPARATION

Practice examinations and simulators are essential to preparing for the CISSP exam because they let you see how well you understand the subject, figure out where you need to improve, and get used to how the exam is set out. This section will discuss many ways to use practice examinations and simulators to help you prepare for the CISSP exam.

- **Use Multiple Sources:** Leverage practice exams and simulators from various sources, such as official (ISC)² materials, reputable third-party providers, and study guides. This will expose you to a diverse range of questions and help you become familiar with different question formats and styles.

- **Simulate Exam Conditions:** To get the most out of practice exams, try to recreate the actual exam conditions as closely as possible. Set a timer, limit distractions, and adhere to the same rules and restrictions as you would during the real exam. This will help you build endurance, improve time management, and reduce exam-day anxiety.

- **Analyze Your Performance:** After completing a practice exam or simulator, thoroughly review your performance. Identify any incorrect answers, and analyze your thought process to understand why you made the mistake. This will help you identify areas that require further study and improvement.

- **Focus on Weak Areas:** As you identify areas of weakness through practice exams, prioritize those domains in your study plan. Spend additional time reviewing the material and working through targeted practice questions to reinforce your understanding and improve your performance in those areas.

- **Review Detailed Explanations:** Many practice exams and simulators provide detailed explanations for each question. Take the time to review these explanations, even for questions you answered correctly. This can help reinforce your understanding of the material and provide additional insights or perspectives.

- **Track Your Progress:** Keep a record of your practice exam scores and track your progress over time. This can help you gauge your overall improvement and identify any persistent areas of weakness. Monitoring your progress can also serve as a motivator, encouraging you to keep studying and refining your knowledge.

- **Space Out Practice Exams:** Avoid taking too many practice exams in a short period. Space them out over the course of your study plan, allowing sufficient time for reviewing and reinforcing the material between exams. This will help prevent burnout and ensure you are fully absorbing and understanding the material.

Using these tips and tricks to prepare for the CISSP test, you can use practice exams and simulators to strengthen your grasp of the subject, find places to improve, and build your confidence for the actual exam. Combining these tactics with other study methods and resources will help you prepare for the CISSP exam in a well-rounded and effective way.

5

TIME MANAGEMENT AND ORGANIZATION

BREAKING DOWN STUDY SESSIONS INTO MANAGEABLE BLOCKS FOR CISSP EXAM PREPARATION

To do well on the CISSP exam, you must manage your time and be organized. Breaking up your study time into smaller chunks will help you stay on task, prevent burnout, and use your study time better. When studying for the CISSP exam, this section will discuss ways to break up your study sessions into manageable chunks.

- **Determine Your Optimal Study Duration:** Identify the length of time you can maintain focus and effectively absorb information. This could range from 25 minutes to an hour, depending on your personal preferences and attention span. Use this duration as a guideline for structuring your study blocks.

- **Use the Pomodoro Technique:** The Pomodoro Technique is a popular time management method that involves breaking work into short, focused intervals (usually 25 minutes) called "Pomodoros," followed by a 5-minute break. After completing four Pomodoros, take a longer break of 15-30 minutes. This technique can help you maintain focus and productivity throughout your study sessions.

- **Schedule Regular Breaks:** Regardless of the study duration you choose, ensure that you schedule regular breaks. Breaks are essential for maintaining your focus and preventing burnout. Use breaks to stretch, take a walk, or engage in a brief non-study-related activity to recharge your mental energy.

- **Plan Your Study Sessions:** At the beginning of each week, create a study schedule that outlines the specific topics or domains you will cover during each study block. This will help you maintain a structured approach to your exam preparation, ensuring that you cover all domains and allocate sufficient time to each topic.

- **Prioritize Difficult Topics:** Allocate more time or additional study blocks to challenging topics or areas of weakness. By focusing on these areas, you can improve your understanding and performance in these crucial aspects of the CISSP exam.

- **Vary Your Study Methods:** To keep your study sessions engaging and effective, incorporate a variety of study methods and resources within each block. For example, you might begin with reading a textbook, followed by watching a video tutorial, and then discussing the material with a study group.

- **Set Study Goals:** For each study block, set specific, achievable goals that relate to the material you are covering. This could include mastering a particular concept, completing a set number of practice questions, or creating a summary of a domain. Setting goals can help you maintain focus and provide a sense of accomplishment throughout your study sessions.

Using these tips for managing your time and being organized, you may split your study sessions into small chunks and prepare for the CISSP exam more efficiently. Combining these tactics with other study methods and resources will help you prepare for the CISSP exam in a well-rounded and effective way.

BALANCING STUDY TIME WITH WORK AND PERSONAL LIFE FOR CISSP EXAM PREPARATION

Many CISSP candidates need help to study, work, and take care of their personal lives simultaneously. Time management and organization are crucial to keeping a good balance and ensuring that studying for an exam goes well. In this section, we'll talk about how to find a good balance between studying for the CISSP exam, work, and your personal life.

- **Create a Realistic Study Plan:** Develop a study plan that takes into account your work schedule, personal commitments, and available study time. Be realistic about the time you can dedicate to studying and set achievable goals that consider all aspects of your life.

- **Prioritize Your Time:** Identify and prioritize your most important tasks, both at work and in your personal life, to ensure that you can focus on your CISSP exam preparation without neglecting other responsibilities. Learn to delegate or say no to less important tasks when necessary.

- **Maximize Your Available Time:** Make the most of your available study time by identifying and capitalizing on periods of the day when you are most alert and focused. This could include early morning sessions, lunch breaks, or evenings after work.

- **Establish Boundaries:** Communicate your study goals and needs to family, friends, and coworkers, and establish boundaries to protect your study time. By setting clear expectations, you can minimize interruptions and distractions during your study sessions.

- **Develop a Consistent Routine:** Establishing a consistent study routine can help you maintain balance and ensure that you allocate sufficient time to both your exam preparation and other aspects of your life. Set aside specific days and times each week for studying and stick to this schedule as much as possible.

- **Be Flexible and Adaptable:** Life can be unpredictable, and you may need to adjust your study plan to accommodate changes in your work or personal life. Be flexible and adaptable, adjusting your study schedule as needed while still maintaining focus on your exam preparation.

- **Take Care of Yourself:** Remember the importance of self-care during your CISSP exam preparation. Ensure you get enough sleep, eat well, and engage in regular physical activity to maintain your overall well-being. Taking care of yourself will enable you to be more focused and productive during your study sessions.

- **Schedule Time for Relaxation and Fun:** It's essential to maintain a balance between study and leisure activities. Schedule time for hobbies, socializing, and relaxation to recharge your mental energy and prevent burnout. Regular breaks from studying can help improve your overall focus and productivity.

Using these tips for managing your time and staying organized, you may prepare for the CISSP exam while caring for your work and personal life. Combining these tactics with other study methods and resources will help you prepare for the CISSP exam in a well-rounded and effective way.

TRACKING PROGRESS AND ADJUSTING THE STUDY PLAN AS NEEDED FOR CISSP EXAM PREPARATION

Keeping track of your progress and changing your study plan as needed is essential to managing your time and staying organized when studying for the CISSP exam. If you monitor your progress, you may see where you need to focus more and make the appropriate changes to your study schedule. When studying for the CISSP exam, this section will discuss ways to track your progress and change your study strategy as needed.

- **Regular Self-Assessment:** Conduct regular self-assessments to evaluate your understanding of the CISSP material. This can include taking practice exams, quizzes, or reviewing flashcards. Use the results of these assessments to identify areas of strength and weakness, and adjust your study plan accordingly.

- **Track Your Progress:** Keep a record of your practice exam scores, completed study materials, and milestones achieved. This will help you gauge your overall improvement and recognize any persistent areas of weakness. Monitoring your progress can also serve as a motivator, encouraging you to continue refining your knowledge.

- **Set SMART Goals:** Establish Specific, Measurable, Achievable, Relevant, and Time-bound (SMART) goals for your CISSP exam preparation. Regularly review and adjust these goals based on your progress and any changes in your work or personal life.

- **Review and Update Your Study Plan:** Periodically review your study plan to ensure it remains relevant and effective. Update your plan as needed, based on your progress, newly identified areas of weakness, or changes in your work or personal life. This will help you maintain a structured approach to your exam preparation, ensuring that you allocate sufficient time to each topic.

- **Seek Feedback:** Engage with study groups, online forums, or mentors to seek feedback on your progress and understanding of the CISSP material. Use this feedback to make informed adjustments to your study plan and further refine your knowledge. Embrace Adaptability: Be prepared to adapt your study plan as needed throughout your exam preparation journey. This may involve reallocating time to specific domains, adjusting your study methods, or rescheduling study sessions due to unforeseen circumstances. Embracing adaptability will help you remain focused and committed to your CISSP exam preparation goals.

- **Reassess Your Time Management:** Regularly reassess your time management strategies to ensure they are effective in balancing your study time with work and personal life. Make necessary adjustments to your routine or study blocks to maintain a healthy balance and ensure efficient use of your study time.

Using these tips for managing your time and being organized, you can keep track of your progress and change your study plan as needed as you prepare for the CISSP exam. Combining these tactics with other study methods and resources will help you prepare for the CISSP exam in a well-rounded and effective way.

UTILIZING PRODUCTIVITY TOOLS AND APPS FOR CISSP EXAM PREPARATION

In the digital world, we live in now, you can use a wide range of productivity tools and apps to assist you in studying for the CISSP exam. Using these tools, you can better organize your study materials, manage your time, and keep track of your progress. In this section, we'll discuss some tools and apps that can help you manage your time and stay organized while studying for the CISSP exam.

- **Task Management Apps:** Task management apps like Todoist, Trello, or Asana can help you create and organize your study tasks, set deadlines, and prioritize your workload. These apps provide a clear overview of your study plan, making it easier to manage and adjust as needed.

- **Calendar Apps:** Use calendar apps like Google Calendar or Microsoft Outlook to schedule your study sessions, set reminders, and plan breaks. By incorporating your study plan into your daily routine, you can ensure consistent progress and maintain a balance between your exam preparation, work, and personal life.

- **Note-taking Apps:** Note-taking apps like Evernote, OneNote, or Notion can help you organize and access your study notes, summaries, and other resources in one central location. These apps also offer features like tagging, searching, and linking, making it easier to navigate and review your notes.

- **Flashcard Apps:** Flashcard apps like Anki or Quizlet can be invaluable for mastering CISSP concepts and terminology. You can create your own digital flashcards or access pre-made decks shared by other users. These apps also allow you to track your progress and customize your study sessions based on your needs.

- **Pomodoro Timer Apps:** Pomodoro timer apps like Focus Keeper or TomatoTimer can help you implement the Pomodoro Technique, a time management method that involves breaking work into short, focused intervals followed by brief breaks. Using these apps can help you maintain focus and productivity during your study sessions.

- **Mind Mapping Tools:** Mind mapping tools like XMind or MindMeister can help you visualize complex concepts, create connections between different topics, and improve your overall understanding of the CISSP material. These tools can be particularly useful for organizing and reviewing your notes.

- **Document Storage and Collaboration:** Cloud storage and collaboration platforms like Google Drive or Dropbox enable you to access your study materials from any device and share resources with study partners or mentors. These platforms can also help you maintain an organized digital workspace for your CISSP exam preparation.

By using these productivity tools and apps to study for the CISSP exam, you may better manage your time and stay organized, speed up the way you study, and make the most of your time. Combining these materials with other study methods and approaches will help you prepare for the CISSP exam in a well-rounded and successful way.

6

STAYING MOTIVATED AND OVERCOMING CHALLENGES

IDENTIFYING AND ADDRESSING COMMON OBSTACLES IN CISSP EXAM PREPARATION

Staying motivated and figuring out how to deal with problems is essential to preparing for the CISSP exam. By recognizing and dealing with typical issues, you can keep your mind on the task at hand, build your confidence, and increase your chances of passing the CISSP exam. This section will discuss ways to find and deal with frequent problems when studying for the CISSP exam.

- **Overcoming Procrastination:** Procrastination can significantly hinder your progress and negatively impact your motivation. To overcome procrastination, break your study sessions into smaller, manageable tasks, and set specific goals and deadlines. Use time management techniques, such as the Pomodoro Technique, to help you stay focused and productive.

- **Managing Stress and Anxiety:** CISSP exam preparation can be stressful and anxiety-inducing. To manage stress and anxiety, practice relaxation techniques like deep breathing, meditation, or progressive muscle relaxation. Ensure that you maintain a healthy lifestyle by getting enough sleep, eating well, and engaging in regular physical activity.

- **Addressing Knowledge Gaps:** It's natural to encounter knowledge gaps while studying for the CISSP exam. Identify these gaps by taking practice exams and engaging in self-assessment. Allocate additional study time to address these areas and seek guidance from mentors, study groups, or online forums.

- **Balancing Work and Personal Life:** Striking a balance between your CISSP exam preparation, work, and personal life can be challenging. Develop a realistic study plan that takes into account your work schedule and personal commitments. Communicate your study goals to your family, friends, and coworkers to ensure their support and understanding.

- **Staying Motivated:** Maintaining motivation throughout your CISSP exam preparation is crucial. Set achievable milestones and celebrate your successes along the way. Stay connected with fellow CISSP candidates through study groups or online forums to share experiences, insights, and encouragement.

- **Overcoming Test-Taking Anxiety:** Test-taking anxiety can negatively affect your performance on the CISSP exam. Practice relaxation techniques, take regular breaks during your study sessions, and familiarize yourself with the exam format and environment to reduce anxiety. Utilize practice exams and simulators to build your test-taking confidence.

- **Adapting to Different Learning Styles:** Recognize that everyone has different learning styles and preferences. Experiment with various study techniques, such as reading, listening to lectures, watching videos, or engaging in hands-on activities, to determine which methods work best for you. Adapt your study plan accordingly to maximize your comprehension and retention of the CISSP material.

Using these tactics and techniques, you can efficiently find and deal with typical problems while preparing for the CISSP exam. Staying motivated and working through issues can help you prepare for the CISSP exam in a well-rounded and successful way. This will increase your chances of passing the exam and advancing your cyber-security career.

DEVELOPING A GROWTH MINDSET FOR CISSP EXAM PREPARATION

A growth attitude is essential to stay motivated and deal with problems as you prepare for the CISSP exam. A growth mindset encourages people to think they can improve their intelligence, skills, and abilities through hard work, tenacity, and learning from mistakes. In this section, we'll talk about developing a growth attitude when studying for the CISSP exam.

- **Embrace Challenges:** Recognize that challenges are opportunities for growth and learning. When you encounter difficult CISSP concepts or tasks, approach them with curiosity and a willingness to learn. By embracing challenges, you will become more resilient and adaptive in your exam preparation journey.

- **Learn from Failure:** Understand that failure is an integral part of the learning process. When you encounter setbacks or make mistakes, view them as valuable learning experiences rather than as indicators of your abilities. Reflect on your failures, identify areas for improvement, and use this knowledge to refine your study strategies.

- **Value Effort:** Appreciate the importance of effort and persistence in achieving your CISSP exam preparation goals. Recognize that success is a result of hard work, dedication, and consistent effort. By valuing effort, you will be more inclined to persevere through challenges and maintain your motivation.

- **Seek Feedback:** Actively seek feedback from mentors, study partners, or online forums to identify areas for growth and improvement. Be open to constructive criticism and use it as an opportunity to refine your knowledge and skills.

- **Cultivate a Positive Attitude:** Maintain a positive attitude towards your CISSP exam preparation by focusing on your progress, celebrating small wins, and acknowledging the effort you have invested in your studies. A positive attitude can help you stay motivated and persevere through challenging moments.

- **Set Realistic Goals:** Establish realistic, achievable goals for your CISSP exam preparation. Break these goals into smaller, manageable tasks and milestones, and track your progress towards achieving them. Setting realistic goals will help you maintain a sense of accomplishment and motivation throughout your study journey.

- **Develop a Support Network:** Connect with fellow CISSP candidates, mentors, or colleagues who share your growth mindset and commitment to continuous improvement. Engage in study groups, online forums, or social media groups to exchange insights, encouragement, and motivation.

- **Emphasize Learning over Performance:** Focus on the learning process rather than solely on your performance in practice exams or other assessment tools. By emphasizing learning, you can develop a deeper understanding of the CISSP material and build a strong foundation for success in the exam and your cybersecurity career.

By developing a growth attitude as you study for the CISSP exam, you can stay motivated, deal with problems, and welcome chances to improve. These tactics and techniques can help you prepare for the CISSP exam in a well-rounded and successful way, putting you on the path to your cybersecurity career goals.

ESTABLISHING A SUPPORT NETWORK FOR CISSP EXAM PREPARATION

A robust support network may help you prepare for the CISSP exam in a big way by giving you encouragement, advice, and inspiration while you study. In this section, we'll talk about why it's essential to form a support network and advise you on how to do so while studying for the CISSP exam.

- **Benefits of a Support Network:** A support network can help you stay motivated, share resources and insights, and troubleshoot challenges you may face during your CISSP exam preparation. Engaging with others who share your goals and aspirations can provide inspiration, encouragement, and a sense of camaraderie.

- **Connect with Fellow CISSP Candidates:** Reach out to other CISSP candidates through study groups, online forums, or social media platforms. These connections can provide valuable insights, tips, and perspectives that can enhance your understanding of the CISSP material and exam strategies.

- **Join or Create a Study Group:** Participating in a study group can help you stay accountable, learn from others, and reinforce your knowledge through discussion and collaboration. Search for local or online CISSP study groups, or consider starting your own with like-minded individuals.

- **Engage with CISSP Mentors:** Seek guidance from experienced professionals who have successfully completed the CISSP exam. CISSP mentors can provide advice on study strategies, exam techniques, and career development. Connect with potential mentors through professional networks, online forums, or industry events.

- **Involve Family and Friends:** Communicate your CISSP exam preparation goals to your family and friends, and ask for their support and understanding. They can provide encouragement, help you maintain a balance between your studies and personal life, and celebrate your achievements along the way.

- **Participate in Online Forums and Communities:** Engage with online CISSP communities, such as Reddit or the (ISC)² Community, to access resources, ask questions, and share your experiences. These forums can be a valuable source of information, support, and motivation throughout your exam preparation journey.

- **Attend Industry Events and Conferences:** Participate in cybersecurity industry events, conferences, and webinars to expand your professional network, stay current with industry trends, and connect with others who share your interest in the CISSP certification.

- **Maintain Your Support Network:** Foster and maintain your support network by regularly engaging with your connections, sharing your progress, and providing support to others. Celebrate your achievements and milestones together and use your collective experiences to overcome challenges and setbacks.

Setting up a support network while studying for the CISSP exam can significantly boost your motivation, confidence, and success. By networking with other candidates, mentors, and experts, you can get helpful tips, information, and encouragement that will help you prepare for the CISSP exam in a well-rounded and effective way.

Rewarding Yourself for Milestones and Accomplishments in CISSP Exam Preparation Rewarding yourself for reaching goals and getting things done as you prepare for the CISSP exam will help you stay motivated, build self-confidence, and keep a positive attitude. In this section, we'll talk about how important it is to recognize and celebrate your successes and give ideas for treating yourself as you study.

- **Importance of Rewarding Yourself:** Acknowledging your hard work, dedication, and progress can help you maintain a sense of accomplishment and motivation throughout your CISSP exam preparation. By rewarding yourself for milestones and accomplishments, you reinforce the positive behaviors and attitudes necessary for success.

- **Set Achievable Milestones:** Break your CISSP exam preparation goals into smaller, manageable milestones. These can include completing a specific number of study hours, mastering a particular domain, or achieving a target score on a practice exam. Setting achievable milestones helps you track your progress and provides opportunities for rewards and celebrations.

- **Choose Meaningful Rewards:** Select rewards that are meaningful, enjoyable, and aligned with your personal values and interests. Examples of rewards can include a relaxing spa day, a favorite meal, a new book or gadget, a weekend getaway, or a night out with friends. Ensure that your rewards are proportional to your accomplishments and do not compromise your study schedule or overall goals.

- **Balance Rewards with Responsibilities:** While rewarding yourself is important, remember to maintain a balance between celebrating your achievements and staying committed to your CISSP exam preparation. Ensure that your rewards do not become distractions or lead to procrastination.

- **Share Your Successes:** Communicate your milestones and accomplishments with your support network, including family, friends, study partners, and mentors. Sharing your successes can help you maintain accountability, receive encouragement, and inspire others in their CISSP exam preparation journey.

- **Reflect on Your Progress:** Regularly reflect on your progress and the milestones you have achieved throughout your CISSP exam preparation. Acknowledging your growth and development can provide motivation, boost your self-confidence, and reinforce your commitment to your goals.

- **Stay Flexible and Adaptive:** Adjust your milestones and rewards as needed to accommodate changes in your study schedule, personal commitments, or progress. Staying flexible and adaptive will help you maintain a positive mindset and ensure that your rewards remain meaningful and motivational.

By recognizing and celebrating your milestones and achievements as you prepare for the CISSP exam, you may stay motivated, keep a positive attitude, and reinforce the behaviors and attitudes you need to succeed. Rewarding yourself for your successes will help you prepare for the CISSP exam in a well-rounded and fun way, increasing your chances of passing the exam and moving up in your cybersecurity career.

7

FINAL EXAM PREPARATION

REVIEWING WEAK AREAS AND GAPS IN KNOWLEDGE FOR CISSP EXAM SUCCESS

Working on your weak spots and knowledge gaps is essential to fully understanding the CISSP material and having the best chance of passing the exam. In this section, we'll talk about how to find and fix your CISSP exam preparation deficiencies and provide advice on improving your knowledge base.

- **Self-Assessment:** Regularly assess your understanding of the CISSP material by taking practice exams, reviewing flashcards, or engaging in discussions with study partners. These activities can help you identify weak areas or gaps in your knowledge that require further attention and study.

- **Analyze Practice Exam Results:** Carefully review your practice exam results to identify areas where you struggled or encountered difficulties. Pay close attention to incorrect answers, and analyze the explanations provided to understand the reasoning behind the correct choices.

- **Create a Gap Analysis:** Develop a gap analysis by listing the CISSP domains and subtopics, and rating your level of confidence or proficiency in each area. Use this analysis to prioritize your study efforts and allocate additional time and resources to weaker areas.

- **Seek Additional Resources:** Utilize additional study resources, such as textbooks, online courses, video tutorials, or articles, to address gaps in your knowledge. Diversify your learning materials to gain a deeper understanding of complex or challenging concepts.

- **Engage with Study Partners or Mentors:** Discuss your weak areas with study partners, mentors, or online communities. These individuals can provide guidance, clarification, or alternative explanations that can help you better understand difficult concepts.

- **Reinforce Learning through Repetition:** Review and reinforce your understanding of weak areas through repetition and active learning techniques. Create flashcards, take notes, or use mnemonic devices to help you memorize key concepts and terminology.

- **Apply Knowledge to Real-World Scenarios:** Enhance your understanding of weak areas by applying your knowledge to real-world scenarios or case studies. This can help you visualize how CISSP concepts are applied in practice and deepen your comprehension of the material.

- **Set Goals and Deadlines:** Establish specific, achievable goals related to improving your weak areas, and set deadlines for achieving these goals. This will help you stay accountable and focused on addressing gaps in your knowledge.

- **Track Progress and Adjust Your Study Plan:** Regularly track your progress in addressing weak areas and gaps in your knowledge. Adjust your study plan as needed to ensure that you are making progress and allocating your time and resources effectively.

During your CISSP exam preparation, you can better comprehend the subject and increase your chances of passing by concentrating on your weak spots and knowledge gaps. Using these tactics and techniques will help you create a strong foundation of knowledge. This will give you the confidence to take the CISSP exam and reach your professional goals in cybersecurity.

TAKING A COMPREHENSIVE PRACTICE EXAM FOR CISSP EXAM SUCCESS

Taking complete practice exams is an essential part of preparing for the CISSP exam, as they give you helpful information about how well you grasp the topic, show you where you need to improve, and get you used to the exam format. In this section, we'll discuss the benefits of taking thorough practice examinations and advise how to get the most out of them while preparing for the CISSP exam.

- **Benefits of Comprehensive Practice Exams:** Practice exams help you gauge your understanding of the CISSP material, identify weak areas, and familiarize yourself with the exam format and question types. Additionally, they can help you build test-taking strategies, manage exam anxiety, and improve your time management skills. Choose High-Quality Practice Exams: Select practice exams from reputable sources, such as the Official (ISC)² CISSP Practice Tests or other well-regarded providers. High-quality practice exams will closely resemble the actual CISSP exam in terms of format, content, and difficulty level.

- **Schedule Practice Exams Throughout Your Study Plan:** Integrate comprehensive practice exams into your study plan at regular intervals. Begin with an initial practice exam to establish a baseline for your knowledge and then schedule additional practice exams to track your progress and ensure that you are improving.

- **Simulate the Actual Exam Environment:** To maximize the effectiveness of your practice exams, create an environment that closely mimics the actual CISSP exam setting. Set a timer, eliminate distractions, and adhere to the same breaks and time constraints as the real exam. This will help you build stamina and develop test-taking strategies.

- **Review Your Results and Analyze Performance:** After completing each practice exam, thoroughly review your results to identify areas that need improvement. Analyze your performance in each domain, and use this information to adjust your study plan and prioritize areas that require additional focus.

- **Learn from Incorrect Answers:** Pay close attention to the questions you answered incorrectly, and review the provided explanations to understand the reasoning behind the correct answers. This will help you identify gaps in your knowledge and avoid making similar mistakes on the actual exam.

- **Develop Test-Taking Strategies:** Use comprehensive practice exams to develop and refine test-taking strategies, such as pacing yourself, eliminating incorrect answer choices, and flagging questions for review. Implementing these strategies can help you manage your time effectively and improve your overall performance on the exam.

- **Build Confidence and Manage Anxiety:** Regularly taking practice exams can help you build confidence in your knowledge and abilities, as well as reduce exam-related anxiety. As you become more familiar with the exam format and content, you will feel better prepared and more at ease when taking the actual CISSP exam.

By preparing for the CISSP exam using full-length practice tests, you can learn how well you comprehend the information, develop test-taking techniques, and feel more confident in your abilities. These tips and tricks will help you get the most out of your practice examinations. This will increase your chances of passing the CISSP exam and help you advance your cybersecurity career.

CAT VS. LINEAR EXAM FORMAT FOR THE CISSP EXAM

As the CISSP exam assesses the candidates' competency, the testing format plays a crucial role. Below are some differences between the Computer Adaptive Testing (CAT) and Linear exam formats for the CISSP exam and discuss the pros and cons of each.

Section 1: Understanding the CAT Exam Format

The CISSP exam transitioned to the CAT format in December 2017. The CAT format is an advanced method of testing that adapts to the test taker's performance during the exam. Here's how it works:

1. The exam begins with a question of moderate difficulty.

2. Based on the test taker's response, the next question is either more or less difficult.

3. The exam continues to adjust question difficulty based on performance.

4. The test concludes when the candidate either achieves a passing score or exhausts the maximum number of questions.

Pros of CAT Exam Format:

- **Tailored to individual ability:** The adaptive nature of the exam ensures that candidates are tested at an appropriate difficulty level, providing a more accurate assessment of their knowledge.

- **Shorter duration:** The CAT exam has a maximum of 150 questions and must be completed within 3 hours, making it a shorter and more time-efficient exam compared to the Linear format.

- **Faster results:** Candidates receive a preliminary pass/fail result immediately upon completing the exam, letting them know their outcome sooner.

Cons of CAT Exam Format:

- **No question review:** Once a candidate answers a question, they cannot go back and review or change their answer.

- **Increased stress:** The adaptive nature of the exam may cause anxiety for some candidates, as they may be uncertain about their performance throughout the test.

Section 2: Understanding the Linear Exam Format

Before implementing the CAT format, the CISSP exam followed a Linear design. In this format, candidates were presented with a fixed number of questions in a pre-determined order, with no adaptation based on their performance. The Linear format had 250 questions and a 6-hour time limit.

Pros of Linear Exam Format:

- **Consistent difficulty:** The questions are presented in a fixed order, ensuring that all candidates face the same difficulty level.

- **Opportunity for review:** Candidates can review and change their answers before submitting the exam, allowing for a more thorough approach to the test.

- Cons of Linear Exam Format:

- **Longer duration:** With 250 questions and a 6-hour time limit, the Linear format is more time-consuming than the CAT format.

- **Less accurate assessment:** The fixed question order may not accurately represent a candidate's knowledge, as it does not adapt to their performance.

The transition from the Linear to the CAT format for the CISSP exam has significantly changed how candidates are assessed. While the CAT format offers a more tailored and time-efficient testing experience, it may also increase stress levels for some candidates. On the other hand, the Linear format provides a consistent level of difficulty and the opportunity to review answers but is more time-consuming and may not provide an accurate assessment of a candidate's knowledge. Understanding the differences between the two formats can help candidates better prepare for the CISSP exam and achieve success on their certification journey.

ENSURING SUCCESS ON THE BIG DAY

Your hard work, devotion, and preparation all come together on the day of your CISSP exam. To ensure you do your best on exam day, following several crucial guidelines and methods is essential. This section will discuss what you can do to do your best on the CISSP exam and stay calm and sure of yourself.

- **Get a Good Night's Sleep:** Adequate rest is crucial for optimal cognitive functioning. Ensure that you get a full night's sleep before your exam day to help you stay alert and focused during the test.

- **Eat a Nutritious Breakfast:** Start your day with a healthy, balanced breakfast to provide the energy and nutrients needed for sustained mental performance. Opt for foods that are high in protein, fiber, and complex carbohydrates to maintain stable blood sugar levels and prevent energy crashes.

- **Arrive Early:** Plan to arrive at the testing center at least 30 minutes before your scheduled exam time. This will give you ample time to check-in, familiarize yourself with the testing environment, and address any last-minute concerns or issues.

- **Bring Necessary Identification and Documentation:** Ensure that you bring the required identification and any necessary documentation, such as your exam confirmation email, to the testing center. Double-check the (ISC)² exam requirements to confirm the specific items needed for your CISSP exam.

- **Dress Comfortably:** Wear comfortable clothing and dress in layers to accommodate fluctuations in room temperature. Being physically comfortable can help you maintain focus and minimize distractions during the exam.

- **Manage Time Effectively:** Be mindful of the time constraints on the CISSP exam, and pace yourself accordingly. Allocate a specific amount of time per question and monitor your progress throughout the exam. If you encounter a particularly challenging question, make an educated guess, mark it for review, and move on to the next question.

- **Utilize Test-Taking Strategies:** Employ the test-taking strategies you developed during your CISSP exam preparation, such as eliminating incorrect answer choices, flagging questions for review, and reading each question carefully to ensure comprehension.

- **Stay Calm and Focused:** Maintain a calm and focused mindset throughout the exam. If you feel overwhelmed or anxious, take a few deep breaths or utilize relaxation techniques to regain your composure.

- **Take Breaks When Needed:** Although the CISSP exam has specific break times, if you feel fatigued or overwhelmed, consider taking a short, unscheduled break to clear your mind and recharge. Use the break to stretch, hydrate, and refocus before resuming the exam.

- **Trust Your Preparation:** Have confidence in the knowledge and skills you've developed during your

CISSP exam preparation. Remember that you have invested significant time and effort into your studies, and trust your ability to succeed on the exam.

Using these methods will help you put your knowledge and abilities to good use, which will increase your chances of passing the CISSP exam and help you move up in your cybersecurity career.

EVALUATING YOUR PERFORMANCE AND MOVING FORWARD

Once you've finished your CISSP exam, you must think about how you did and figure out the following stages in your cybersecurity career. In this section, we'll talk about how to evaluate yourself after the CISSP exam, how to analyze your exam results, and what to do next.

- **Understanding Your Exam Results:** After completing the CISSP exam, you will receive a preliminary pass or fail result. The official exam results will be emailed to you by (ISC)² within 2-5 business days. The email will contain a detailed score report, which provides your scaled score and the passing scaled score. Use this information to assess your performance and identify areas for improvement if needed.

- **Passing the CISSP Exam:** If you have successfully passed the CISSP exam, congratulations! Your next step is to complete the (ISC)² endorsement process, which verifies your work experience and validates your qualifications for the CISSP certification. Once your endorsement is approved, you will receive your CISSP certification and become a member of the (ISC)^2 community.

- **Failing the CISSP Exam:** If you did not pass the CISSP exam, don't be discouraged. Use this experience as an opportunity to learn and improve. Review your score report and identify the domains in which you performed poorly. Focus on these areas and develop a plan to address the gaps in your knowledge and understanding.

- **Reassess Your Study Plan:** Reevaluate your study plan and determine if any changes are needed. Consider incorporating new resources, adjusting your study schedule, or seeking additional support from study groups or mentors. Remember that effective CISSP exam preparation requires a combination of resources, strategies, and techniques.

- **Retaking the CISSP Exam:** You can retake the CISSP exam after a waiting period of 30 days. Keep in mind that you are allowed a maximum of three attempts within a 12-month period. Use the time between exam attempts to refine your study plan, address weak areas, and continue building your knowledge and skills.

- **Continuous Professional Development:** Once you have obtained your CISSP certification, it is essential to engage in continuous professional development to maintain your certification and stay current in the ever-evolving field of cybersecurity. Participate in industry events, attend conferences, and seek additional training opportunities to expand your knowledge and stay up-to-date with the latest trends and developments.

- **Networking and Community Involvement:** Connect with other cybersecurity professionals and become involved in the (ISC)² community. Networking can open doors to new career opportunities, provide valuable insights, and help you stay informed about the latest advancements in the industry.

- **Pursuing Additional Certifications:** Consider pursuing additional certifications or specializations to further enhance your skills and expertise in the cybersecurity field. Advanced certifications, such as the (ISC)² CCSP (Certified Cloud Security Professional) or the ISACA CISM (Certified Information Security Manager), can help you advance your career and demonstrate your commitment to professional growth.

By doing these procedures after the exam, you can evaluate how well you did on the CISSP exam and go further in cybersecurity. Whether you passed or failed the test, remember that you may invariably learn and progress. Invest in your professional growth, learn more, and build your network to succeed in cybersecurity in the long run.

CONCLUSION

EMBRACING THE JOURNEY WITH AN EFFECTIVE CISSP STUDY PLAN

As we near the end of this book, it is essential to review the most critical parts of a good CISSP study plan and stress how important it is to prepare for the exam in a thorough and organized way. Getting the CISSP certification is challenging but gratifying because it can help you open new doors and advance in your cybersecurity profession.

Throughout this ebook, we have explored various components of a successful CISSP study plan, including:

- **Assessing your current knowledge and experience**

- **Determining your learning style**

- **Setting realistic goals and deadlines**

- **Creating a personalized study schedule**

- **Gaining an in-depth understanding of the CISSP exam domains**

- **Utilizing diverse study materials and resources**

- **Employing active learning strategies and techniques**

- **Managing your time effectively and maintaining a balance between study, work, and personal life**

- **Staying motivated and overcoming challenges during the preparation process**

- **Preparing for the exam day and executing your test-taking strategies**

A good CISSP study strategy only works for some. Instead, it is a personalized, changing plan that changes as you go and considers your particular learning style, skills, and shortcomings. As you start this path, remember that the keys to success on the CISSP exam and cybersecurity are tenacity, devotion, and constant learning.

FOSTERING SUCCESS ON THE CISSP EXAM WITH CONFIDENCE AND ENCOURAGEMENT

As we wrap up this book, it's important to remember how important it is to give yourself confidence and support as you study for the CISSP exam. The path to CISSP certification may seem complicated, but if you have the correct attitude, are dedicated, and don't give up, you may be successful and take your cybersecurity career to new heights.

Throughout this booklet, we've given you a detailed plan to help you prepare for the CISSP exam. It's important to realize that your success on the test doesn't just depend on what you study or how you study. It also depends on your belief in yourself and how well you can handle problems. To ensure your success on the CISSP exam, keep the following key points in mind:

- **Believe in Yourself:** Have faith in your abilities and trust the hard work you have invested in your preparation. Recognize that you are capable of achieving your goals and that the CISSP certification is within your reach.

- **Embrace the Learning Process:** View your CISSP exam preparation as an opportunity for personal and professional growth. Focus on building a strong foundation of knowledge and skills that will not only help you pass the exam but also serve you well throughout your cybersecurity career.

- **Stay Persistent:** Understand that setbacks and challenges are a natural part of the learning process. Stay persistent and committed to your goals, and remember that every step you take, no matter how small, brings you closer to success.

- **Maintain a Positive Attitude:** A positive mindset can make all the difference in your exam preparation. Stay optimistic and view challenges as opportunities to learn and grow. Celebrate your achievements, no matter how small, and use them as motivation to keep pushing forward.

- **Seek Support:** Surround yourself with a network of supportive individuals, such as mentors, peers, and study groups, who can offer guidance, encouragement, and camaraderie throughout your CISSP journey. Leverage their collective wisdom and experience to enhance your own learning and development.

- **Stay Curious:** Keep the fire of curiosity alive and nurture your passion for learning. Stay up-to-date with the latest developments in the cybersecurity field, and never stop exploring new ideas, concepts, and technologies.

Your path to CISSP certification may be challenging, but if you have the correct attitude and keep working hard, you can get beyond any problems. As you study for the CISSP exam, remember you are not alone. Many experts have been in your shoes and successfully made it through this journey. You can join them with determination, patience, and a good attitude.

Accept the challenge, have faith in yourself, and keep your goals in mind. You can pass the CISSP exam, and we wish you the best as you move forward in your cybersecurity career.

Significance of Continuous Learning and Professional Development in the Cybersecurity Field

As we wrap up this ebook, it's important to stress how important it is to keep learning and improving your skills in the ever-changing world of cybersecurity. Getting CISSP certified starts a lifelong quest for education, growth, and greatness in cybersecurity.

Cybersecurity is constantly changing, so experts must keep up with the newest trends, threats, and technologies. In this situation, continuous learning and professional development aren't just nice; they must stay ahead of the competition and keep the companies you work for safe and robust.

Here are some key points to consider as you embark on your path of lifelong learning in the cybersecurity field:

- **Keep Your Skills Up-to-Date:** The cybersecurity landscape is constantly changing, with new threats, vulnerabilities, and technologies emerging regularly. Stay abreast of the latest developments and ensure your skills remain relevant and effective.

- **Pursue Advanced Certifications:** After obtaining the CISSP certification, consider pursuing additional specialized certifications to deepen your knowledge in specific areas of cybersecurity. Advanced certifications can help you advance your career and demonstrate your commitment to professional growth.

- **Attend Conferences and Workshops:** Participate in industry conferences, workshops, and webinars to stay informed about the latest advancements, trends, and best practices in cybersecurity. Networking with peers and experts can also provide valuable insights and opportunities for collaboration.

- **Engage with Professional Organizations:** Become an active member of professional cybersecurity organizations, such as (ISC)², ISACA, and SANS, to gain access to exclusive resources, networking opportunities, and knowledge-sharing platforms.

- **Pursue Further Education:** Consider enrolling in graduate or postgraduate programs in cybersecurity or related fields to expand your knowledge and enhance your credentials. Formal education can provide a structured approach to learning and help you develop a deep understanding of complex cybersecurity topics.

- **Share Your Knowledge:** Give back to the cybersecurity community by sharing your knowledge and expertise. Write articles, speak at conferences, or mentor others to contribute to the collective wisdom of the profession.

- **Embrace a Growth Mindset:** Cultivate a growth mindset that embraces challenges and sees setbacks as opportunities for learning and improvement. This mindset will empower you to continually push your boundaries and strive for excellence in your professional life.

In conclusion, the CISSP certification is essential to becoming a successful cybersecurity expert. But it's crucial to note that certification is not the end of the quest for information, progress, and quality. Accept ongoing learning and professional development as essential parts of your cybersecurity job, and make a promise to be up-to-date, practical, and effective in a threat landscape that is constantly evolving.

By committing to lifelong learning and professional progress, you can ensure that you will continue to be successful in cybersecurity and help protect the digital assets and information that are the foundation of our modern civilization. So, keep asking questions, getting involved, and working hard to make a difference in cybersecurity.

Enjoy the journey, stay committed to your goals, and remember that the knowledge and abilities you gain will help you pass the CISSP exam and be helpful in your professional life. Good luck, and I hope you do well!

ABOUT THE AUTHOR

Dr. Leonard Simon, a distinguished figure in the field of cybersecurity, has been an influential presence in the IT industry since 2000. With a career spanning over two decades, Dr. Simon has made significant contributions to the cybersecurity field and played a vital role in shaping the next generation of cybersecurity professionals through his work as an adjunct professor.

Dr. Simon's academic journey began at Florida International University, where he earned a Bachelor's degree in Information Technology. His passion for cybersecurity and information systems led him to pursue a Master's in Management Information Systems and Security from Nova Southeastern University. His thirst for knowledge continued as he obtained his Doctorate in Information Assurance and Cybersecurity from Capella University, positioning him as an expert.

After a decade of experience in the IT industry, Dr. Simon transitioned to cybersecurity in 2010, where he assumed various roles in both the public and private sectors. His expertise in information assurance, risk management, and network security has been instrumental in helping organizations develop robust cybersecurity strategies and defend against emerging threats.

In 2012, Dr. Simon began sharing his extensive knowledge and experience with students as an adjunct professor at several prestigious universities nationwide. His teaching repertoire includes various cybersecurity courses, such as network security, ethical hacking, digital forensics, and cybercrime. Dr. Simon's dedication to his students and ability to convey complex concepts engagingly have earned him a reputation as an exceptional educator.

Dr. Leonard Simon's extensive experience in the cybersecurity field, combined with his commitment to teaching, has made him an invaluable resource for organizations and students. As a sought-after speaker at industry conferences and seminars, he continues to share his insights and knowledge with professionals worldwide.

Outside of his professional engagements, Dr. Simon is an active member of several cybersecurity organizations and contributes to industry research and publications. His work has been recognized with numerous awards and accolades, further solidifying his standing as a thought leader in the cybersecurity domain.

Dr. Simon's passion for cybersecurity and dedication to education inspires countless professionals and students to pursue careers in this ever-evolving field. His ongoing commitment to knowledge-sharing and tireless work to improve cybersecurity practices have had a lasting impact on the industry, ensuring a more secure digital landscape for future generations.

WWW.PROFESSORSIMON.COM

email: info@professorsimon.com